Pg. 189   1/24

DIY SCIENCE FAIR FUN!

METEOROLOGY
PROJECT YOUR WAY

    Megan Borgert-Spaniol

Super Sandcastle

An Imprint of Abdo Publishing
abdobooks.com

abdobooks.com

Printed in the United States of America, North Mankato, Minnesota

102023
012024

Design: Aruna Rangarajan, Mighty Media, Inc.
Production: Mighty Media, Inc.
Editor: Liz Salzmann
Cover Photographs: Adobe Stock; Mighty Media, Inc.
Interior Photographs: Adobe Stock, pp. 1 (anemometer), 11 (sun, beach), 15, 24, 28 (beaches, sun); iStockphoto, pp. 4 (both), 6, 10, 14 (water, tray), 17 (step 5), 26, 31; Mighty Media, Inc., pp. 9 (thermometer), 14 (cups, sand, thermometers), 16 (experiment), 17 (step 4), 18, 21 (pebbles, sand), 28 (experiment), 29 (experiment); NASA, p. 5; Shutterstock, pp. 7, 8, 9 (girl), 11 (girl), 13, 14 (stopwatch), 16 (thermometer), 19, 22, 25, 27, 29, 30
Design Elements: Shutterstock

Library of Congress Control Number: 2023939471

Publisher's Cataloging-in-Publication Data
Names: Borgert-Spaniol, Megan, author.
Title: Meteorology project your way / by Megan Borgert-Spaniol
Description: Minneapolis, Minnesota : Abdo Publishing, 2024 | Series: DIY science fair fun! | Includes online resources and index.
Identifiers: ISBN 9781098292065 (lib. bdg.) | ISBN 9781098278960 (ebook)
Subjects: LCSH: Do-it-yourself work--Juvenile literature. | Meteorology--Juvenile literature. | Weather--Juvenile literature. | Science projects--Juvenile literature. | Science fair projects--Juvenile literature.
Classification: DDC 507.8--dc23

Super SandCastle™ books are created by a team of professional educators, reading specialists, and content developers around five essential components—phonemic awareness, phonics, vocabulary, text comprehension, and fluency—to assist young readers as they develop reading skills and strategies and increase their general knowledge. All books are written, reviewed, and leveled for guided reading, early reading intervention, and Accelerated Reader™ programs for use in shared, guided, and independent reading and writing activities to support a balanced approach to literacy instruction.

CONTENTS

EXPLORE METEOROLOGY

Do you love to watch clouds form? Do you wonder what causes storms? You might enjoy meteorology! Meteorology is the study of Earth's atmosphere and weather. Scientists who study meteorology are called meteorologists.

Meteorologists use weather balloons to collect data miles above the ground.

Weather stations have instruments that track weather at Earth's surface.

Meteorologists study the conditions that cause weather. They collect weather data, such as temperature and wind speed. They use the data to forecast the weather.

Satellites show scientists images of Earth's weather from outer space.

BECOME A SCIENTIST!

Scientists use a process called the scientific method. Check out the steps on the next page. You will use this method to **design** your own meteorology project!

THE SCIENTIFIC METHOD

1 ASK A QUESTION
What would you like to find out?

2 GATHER INFORMATION
What information do you need to understand your topic?

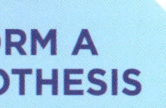

3 FORM A HYPOTHESIS
What do you think is the answer to your question?

4 EXPERIMENT
How can you test your hypothesis to find out if it is correct?

5 RECORD THE RESULTS
What did you observe in your experiment?

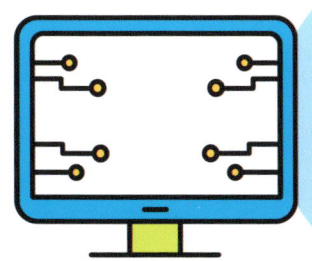

6 WRITE A CONCLUSION
Did your results support your hypothesis?

ASK A QUESTION

What topic do you want to learn about?

Maybe you are interested in wind. Or maybe you want to know more about temperature. Start asking questions! Write your questions in a notebook so you don't forget them.

What causes wind?

What makes wind blow in different directions?

8

How does
the temperature
change during
the day?

Why is it
hotter in some
places than
in others?

GATHER INFORMATION

Maybe you have a lot of questions. That's great!

Scientists often have many questions they'd like to answer. But for now, choose one to **focus** on. Save the others for **future** projects.

It's time to **research** your **topic**. You can gather **information** from many different sources.

Read online articles about the topic.

Read books about the topic.

Talk to scientists or other experts.

What did you learn about your **topic**? Write it down in your notebook. Then you'll have all the **information** you need in one place.

The temperature of an area partly depends on how much heat it gets from the sun.

An area's temperature also depends on how far it is from the ocean.

Areas near the ocean are cooler in summer and warmer in winter than inland areas.

STEP 3

FORM A HYPOTHESIS

After you research your topic, it's time to form a hypothesis.

Your hypothesis is what you believe is the answer to your question. First, revisit your question. Do you want to change it based on what you learned? Then think of a few different hypotheses. Record them all in your notebook.

QUESTION: Does the sun's heat warm up land and water at the same rate?

HYPOTHESIS 1

Land warms up faster than water.

HYPOTHESIS 2

Water warms up faster than land.

HYPOTHESIS 3

Land and water warm up at the same rate.

Now, choose which hypothesis makes the most sense based on your **research**.

I learned that areas near the ocean are cooler in summer.

So, I think land warms up faster than water.

13

PREPARE YOUR LAB

Get ready to test your hypothesis.

Find an area with a sturdy table or counter to work on. Then gather the supplies you'll need for your science experiment.

SUPPLIES

2 identical
plastic cups

sand

water

tray

stopwatch
or clock

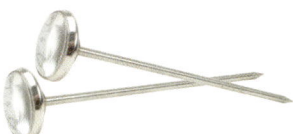

2 thermometers

LAB RULES

All labs have rules that scientists have to follow. Here are some rules for your lab. They will help you stay safe and have fun while doing your experiment!

→ **Ask an adult** for permission to use the materials and do the experiment.

→ **Ask for help** with sharp or hot tools.

→ **Wear goggles** and gloves to protect your eyes and hands.

→ **Clean up** when you are done and put everything away.

STEP 4

EXPERIMENT!

You've gathered the supplies. You've prepared the lab. It's time to experiment!

1 Fill one of the cups most of the way with the sand. This represents land.

2 Fill the other cup to the same level with water. This represents the ocean.

3 Place a thermometer in each cup. Put the cups on a tray.

4 Set the tray outside in the sun. Make sure it is in a safe spot away from any shade.

5 Record the starting temperatures of the water and sand. Check the thermometers every 5 minutes for the next 30 minutes. Each time, record the temperatures and how many minutes have passed.

LAB TIP

Check the weather forecast before doing this experiment. Choose a day that's likely to be sunny.

Look at the results of your experiment so far. You might be ready to draw a conclusion. But first, consider any other **variables** that might affect your results.

Some colors and materials **absorb** more heat than others. You used identical cups so they would absorb the same amount of heat.

+8°

+16°

You filled both cups to the same height. So, the water and sand took up the same amount of space.

I also put both cups in the same spot. So, they got the same amount of heat from the sun.

This means the only **variable** was the material inside the cups.

19

RECORD THE RESULTS

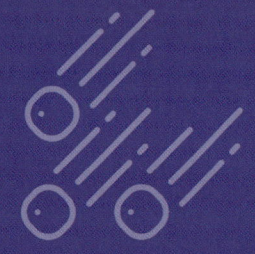

During experiments, scientists record data and other observations. You wrote down the temperature of each material every five minutes. Now, it's time to record your data to share with others.

Scientists often use tables and graphs. This helps make the results easy for others to understand. A table organizes **information** in rows and columns.

WATER		
MINUTES PASSED	TEMPERATURE (°F)	DEGREE INCREASE
0	82	0
5	85	3
10	86	4
15	88	6
20	89	7
25	90	8
30	90	8

SAND		
MINUTES PASSED	TEMPERATURE (°F)	DEGREE INCREASE
0	92	0
5	95	3
10	98	6
15	100	8
20	103	11
25	105	13
30	108	16

A line graph shows the relationship between two **variables**.

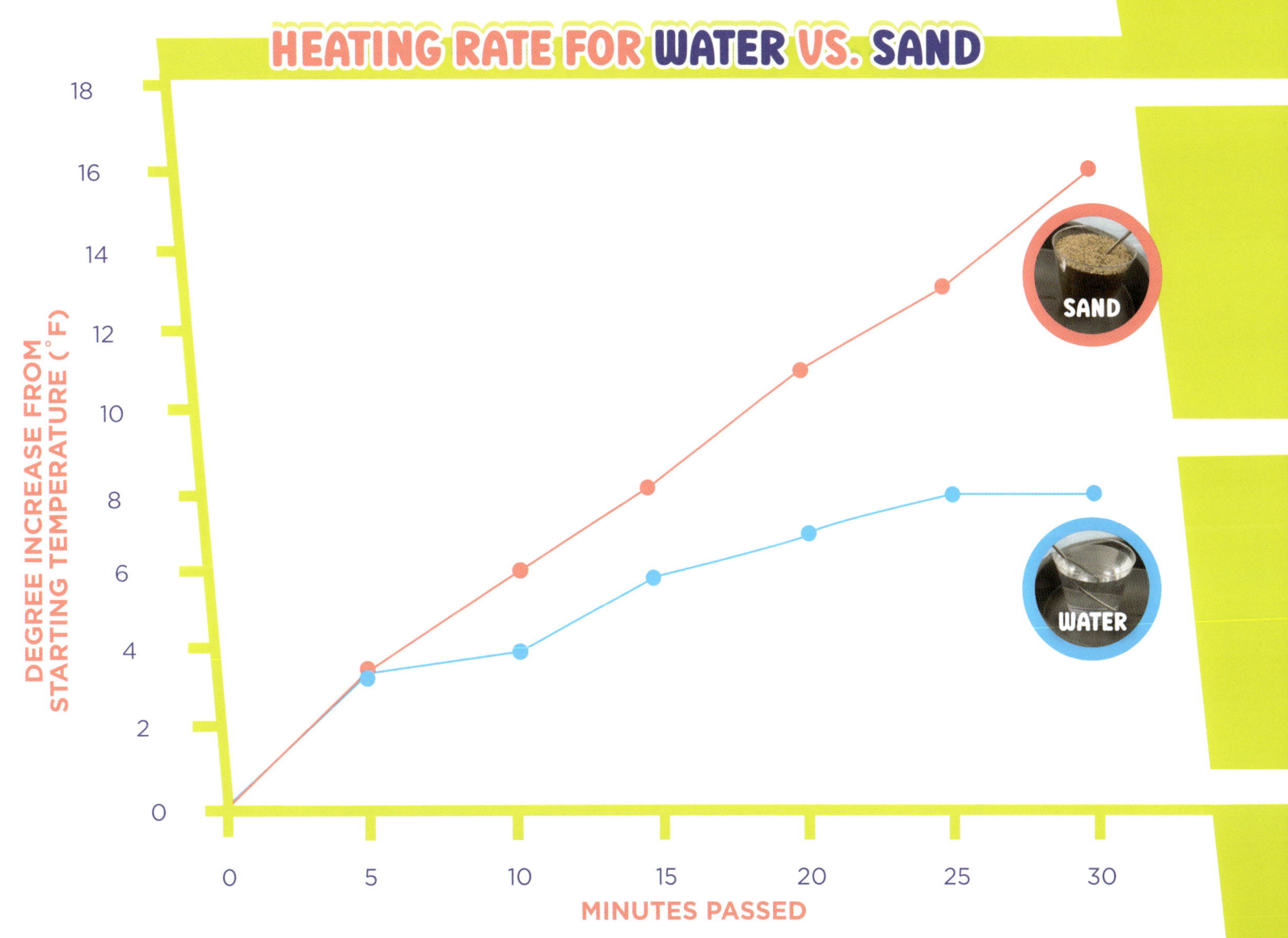

HEATING RATE FOR WATER VS. SAND

DEGREE INCREASE FROM STARTING TEMPERATURE (°F)

MINUTES PASSED

SAND

WATER

WRITE A CONCLUSION

You recorded your results. Now it's time to write your conclusion. This is a **summary** of your experiment. Your conclusion provides the answer to your original question. It also states whether your results support your hypothesis.

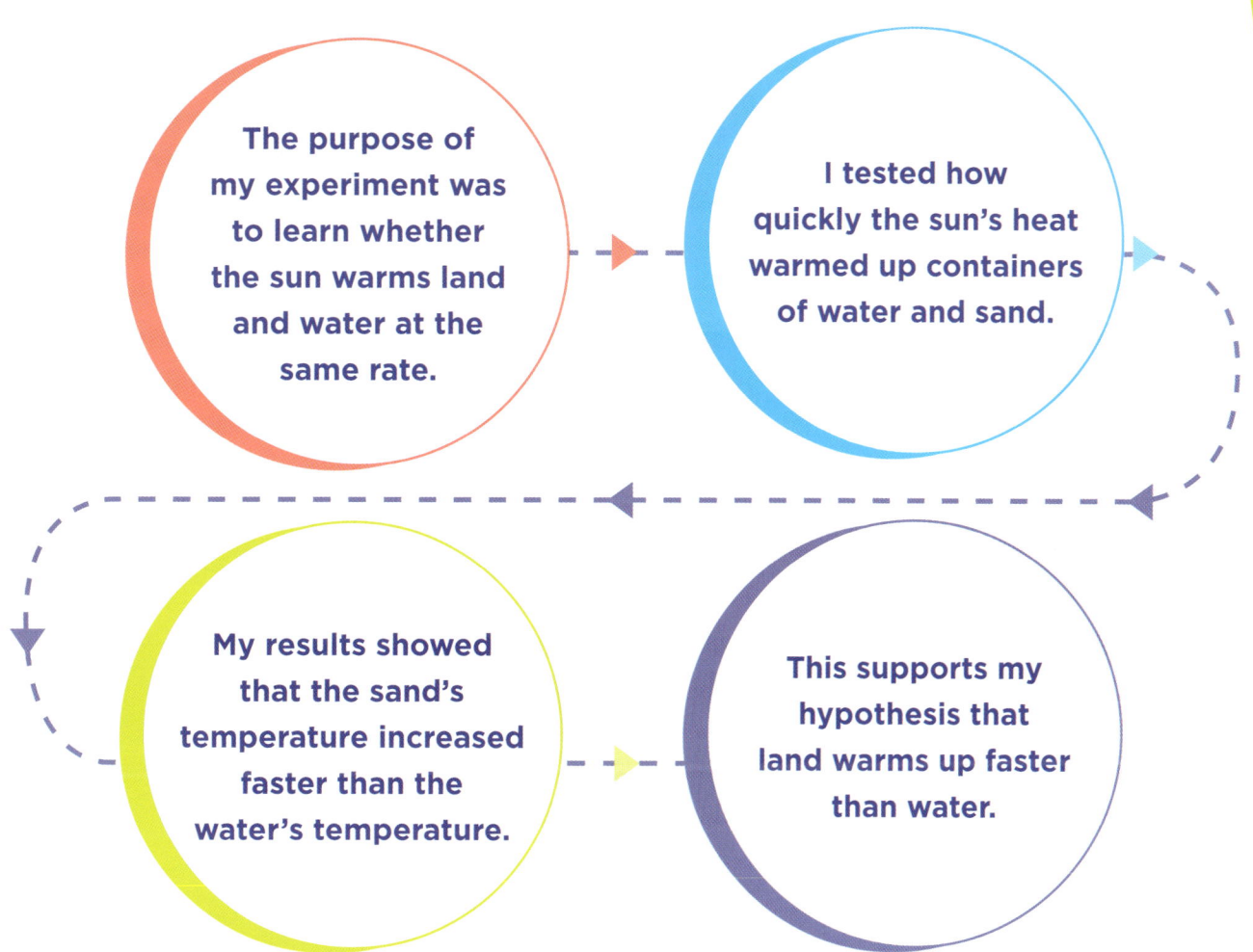

The purpose of my experiment was to learn whether the sun warms land and water at the same rate.

I tested how quickly the sun's heat warmed up containers of water and sand.

My results showed that the sand's temperature increased faster than the water's temperature.

This supports my hypothesis that land warms up faster than water.

Many scientists find that their hypotheses were wrong. There is nothing wrong with being wrong! It means you did your experiment without **bias** and were surprised by the results. That is another mark of a great scientist!

FURTHER RESEARCH

A conclusion offers an answer to your original question. But it can bring up new questions too! For scientists, the end of one experiment often leads to more **research** and new hypotheses to be tested.

What new questions do you have? What could you research further? Do you have a new hypothesis to test?

How do clouds affect the temperature at Earth's surface?

Do land and water cool off at different rates?

25

PRESENT YOUR PROJECT

Young scientists share their **research** at science fairs. Students share what they learned with classmates, teachers, parents, and sometimes judges. It is a chance to show all the work they put into their experiments.

THERE ARE LOTS OF FUN WAYS TO SHARE YOUR NEW KNOWLEDGE!

Demonstrate or show a video of your experiment.

Create comics or other drawings to show your project in a fun way.

Include props, models, or dioramas.

One way to present a project is with a display board. It should show how you followed the scientific method. Turn the page to see a display board of the project in this book!

QUESTION

Does the sun's heat warm up land and water at the same rate?

RESEARCH

The temperature of an area partly depends on how much heat it gets from the sun. It can also depend on distance from the ocean. Areas near the ocean are cooler in summer and warmer in winter than inland areas.

HYPOTHESIS

I think land warms up faster than water.

EXPERIMENT

In my experiment, I put cups of water and sand in the sun. I wrote down the temperature of each one every five minutes for half an hour.

Water

Sand

RESULTS

My results showed that the sand's temperature increased faster than the water's temperature.

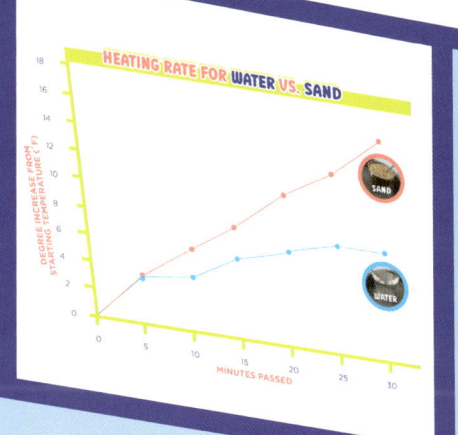

HEATING RATE FOR WATER VS. SAND

SAND

WATER

DEGREE INCREASE FROM STARTING TEMPERATURE (°F)

18
16
14
12
10
8
6
4
2
0

0   5   10   15   20   25   30

MINUTES PASSED

CONCLUSION

The results support my hypothesis that land warms up faster than water.

KEEP ASKING QUESTIONS

Your science project is over. You packed away your display. But don't stop asking questions! What might you do differently if you did the project again? What additional **research** could you do? Is there a related **topic** you would like to explore?

30

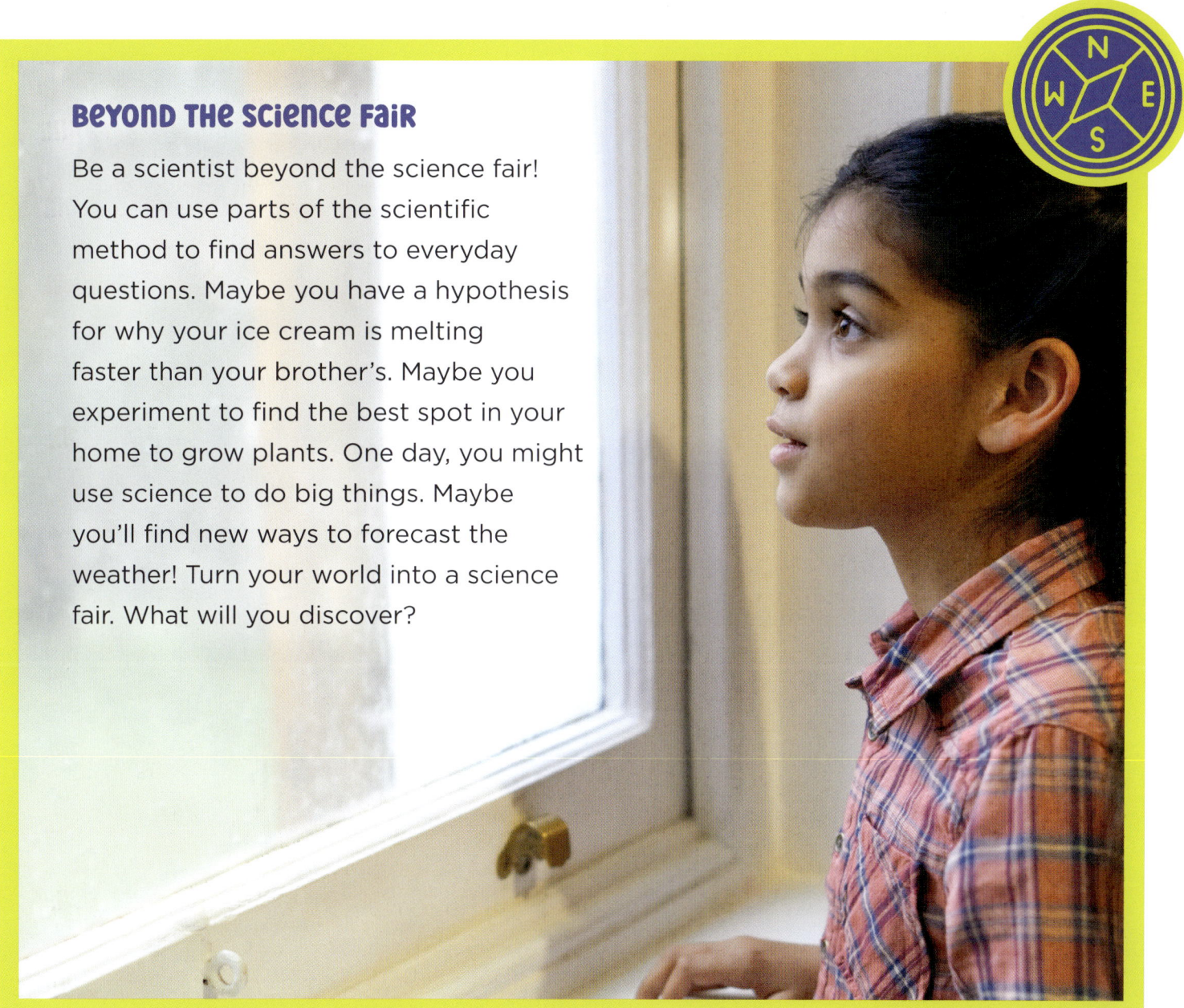

Beyond the Science Fair

Be a scientist beyond the science fair! You can use parts of the scientific method to find answers to everyday questions. Maybe you have a hypothesis for why your ice cream is melting faster than your brother's. Maybe you experiment to find the best spot in your home to grow plants. One day, you might use science to do big things. Maybe you'll find new ways to forecast the weather! Turn your world into a science fair. What will you discover?

GLOSSARY

absorb—to soak up or take in.

bias—showing a preference for one result over another.

design—to plan how something will appear or work.

focus—to concentrate on or pay particular attention to.

future—the time that hasn't happened yet.

information—the facts known about an event or subject.

research—to find out more about something. Also, a study of something to learn new information.

summary—a short statement of the main points.

topic—the main idea or subject.

variable—a factor in a scientific experiment that may change.